El Teide

By Pia Lord

Other Books by Pia Lord

Tenerife Canary Islands Spain

Return to the Sea: A Guide to Open Water Swimming in Mallorca, Spain

The Upper Limit

Let's Take a Trip in Our Spaceship

Aggity, Biggity, Ciggity

The Day the Sun Went Out

The Night the Moon Went Out

Cato the Caterpillar

The Adventures of M.M., Music Mouse

Cat-skills in the Rocky Mountains

Just Pia!

Rhapsody

The A, B, C Collection

Harvest While the Orchard is Aplenty

Foreword

The earth sings at times, but at others it spews forth its rebirth in the recreation of itself and the maintenance of its balance in the universe. Our Earth must know what it is doing. It is alive clearly, yet it is not our doing that it is so. Forces of nature and of God are present that man must observe and understand if we are to harmonize with the universe. Many species go extinct for the very reason that they clash to severely with their environment. Keeping a finely tuned ear and highly observant eye to the processes going on underneath our homes is a good first step to maintaining the survival of our species.

Acknowledgements

I would like to thank my family and friends, for their support in my travels. With such caring, I am able to continue writing books on interesting scientific subjects. It is a privilege to travel and see the world in its rich topographic, geologic, atmospheric and oceanographic history. I especially would like to thank my husband and son for their cooperation for the weeks that I am away. They continue to excel in their endeavors so I know all is well on the home front. When I am away and am concerned for them, I am grateful to the technology that we have which allows us to communicate, seek, find, ask and receive the help we need. I would also like to thank Mr. Fredrik von Weisz, M.S. in Geology for his assistance with the geological information on volcanos. He provided an informative oral narrative on the history and formation of El Teide while exploring on the volcano, as well as assistance in reviewing the manuscript for scientific accuracy. Finally, I would like to thank God for the strength and fortitude given to me to carry out these trips. The presence of the holy spirit is a constant source of comfort.

EL TEIDE

Tenerife Volcano

Canary Islands, Spain

by

Pia Lord

El Teide is the third highest volcano in the world when measured from the sea floor to its peak.

The Canary Islands lie northwest of Africa, off the coast of Morocco and south of the Azores Islands in the North Atlantic Ocean. There are seven main volcanic islands which form the Canary Islands archipelago. They are Tenerife, Fuerteventura, Las Palmas, Gran Canarias, Lanzarote, Hierro, and La Gomera. Tenerife is the largest of the Canary Islands. El Teide is the name given to the volcano which majestically rises from the island of Tenerife. Its peak is at 3,718 meters (12,198 feet) in elevation. Along with being the third highest of all volcanoes in the world, when measured from the ocean floor, it is the highest peak in Spain. El Teide shows *strombolian* and *plinian characteristics*, based on the stratovolcanoes Mount Stromboli and Mount Vesuvius in Italy. In an explosive eruption style, it throws material up into the atmosphere. When erupting, it is a *fumarolic,* steaming, gaseous volcano. Although dormant, it is yet considered an active volcano due to the minor earthquakes and rumblings that scientists document with seismographs between Mount Teide and Santiago del Teide. El Teide has a massive caldera, La Caldera de Las Cañadas, with a secondary smaller peak inside. La Caldera de Las Cañadas is 10.5 miles long, 47-50 miles wide, with walls 1,640 feet high. This caldera has recently been determined to have been formed by plate tectonic shifts 160,000 years ago, rather than from the collapse of the material above the magma chamber. The volcano then began forming 120,000 years ago. The chemistry of the magma is mainly mafic in character. This means it has a lower Silicon Dioxide (SiO_2) content. Rocks that are found on the volcano are basalts.

The island of Tenerife started forming more than 12 million years ago with the last volcanic eruption occurring in 1909. The island has been build up sequentially in 4-5 stages. The peak today is less than 200, 000 years old. Although the volcano has erupted more than 43 times in history according to the Smithsonian Institution, eruptions happened only five times since the 1492. As Pangea split up into several continents, similar to East Africa breaking off the main continent along the East African Rift Valley, volcanic activity and plate tectonic shifts of the Earth formed islands and continents that we know today, including Africa and the Americas. The Canary Islands, an intraplate archipelago formed in the middle of the African tectonic plate. The Earth's internal heating activity caused hotspots under the ocean. The mantel then cracked through weaknesses in the lithosphere, the upper crust. Through these weaker mantel areas, *magma* erupted to the surface, as molten lava. Continually shifting plates, earthquakes, landslides, eruptions, layering of lava and lifting of the lithosphere over time formed these oceanic islands.

All around Tenerife are evidences of past volcanic activity. The topography of the island shows different ages due to the varying landscape. There is caldera, reddish black lava, A'a jagged chunky lava, forest canopy, smooth water eroded volcanic boulders, black obsidian glass volcanic rock formations, and viscous layered lava channels. Atmospheric effects of cloud line and sun glint are also in evidence. A solar effect, sun glint, can be seen over the North Atlantic Ocean from high up on the volcano El Teide, while facing south looking out towards La Gomera. Could sun glint have any effect on volcanic activity? Would temperatures of a hotspot under the sea be increased by the presence of a continual sun glint? Or is sun glint merely a visual phenomenon?

La Caldera de Las Cañadas

Standing high up near the peak of the volcano and looking down, there is an obvious caldera formation visible. The edges of the bowl look somewhat like formations in the Grand Canyon in Arizona. Theories of caldera formation include that volcanic eruptions pushed one area of the earth up to the peak, causing crumbling and erosion around the outside edge. However, more recent scientific studies show that a shifting of the underlying tectonic plates caused the caldera. Landslides occurred forming valleys and rifts, and then later the volcano was formed.

The edges of La Caldera de Las Cañadas and Grand Canyon look alike as layered formations. Lava flow on lava flow, formed the layered lava channels seen in the Masca Gorge, on the southwestern side of the island as well. In the Grand Canyon, sandy material was deposited in a sea and formed massive layered beds. Weathering may be influential on Teide caldera, as it is very significant for the Grand Canyon. However, geological material is deposited in layers during many formational processes.

El Chinyero

Reddish black pahoehoe lava flow (background) and A'a chunky lava (foreground)

Look closely at the center of the above reddish black dark lava flow to see the secondary eruption point. The reddish black lava is smooth or wrinkly. This type is known as pahoehoe. The foreground jagged lava is the a'a type. This area is known as the El Chinyero vent on the northwestern Santiago *rift*. The weaknesses in the mantel become evident each time the volcano erupts. El Teide is dubbed a *decade volcano* due to its history of destructive eruptions. El Teide is close to several towns in Tenerife. The names of these towns are Icod de Los Vinos, Garachico, and Puerto de La Cruz. The people in these towns live literally in the shadow of the volcano. How would you feel living there? Would you like to live on the Island of Tenerife, or any of the volcanic Canary Islands?

Jagged chunks of boulders of volcanic lava rock on the slope of the volcano

In driving up the volcano access road, most of the landscape is filled with this broken volcanic rock or A'a, name in Hawaiian. These are materials from the eruption or weathered broken off pieces. Large chunks are laying all over the place! They are hardened from the cooling process that takes place days or weeks after the eruption. In some areas the local people are squaring off these large chunks and building huts and other small structures. This chunky, jagged soil area is below the forest canopy.

Forest Canopy

Forest canopy on the volcano slope   Pine cone from forest trees

In ascending the slope of the volcano, driving on the mountain road TF-24, there is the alpine region of the forest. The elevation is about 2300 meters. There are numerous trees of the coniferous type. These trees have large pine cones that stand the height of the human adult hand and have a circumference the size of the human adult palm. These trees grow in the lava soil of the landscape, beginning the process of vegetation. Once they have matured to greater heights they produce the large pinecones.

Volcanic Boulders

Smooth *volcanic boulders* show erosion and weathering of rocks

The smooth boulders are found down by the coastline. These were likely thrown by the force of the eruption to the edge of the emerging island. Over millions of years these boulders have been washed by the sands and the water. The constant washing by the sea, and weathering by the rains and winds have brought them to a state in which they have this smooth appearance. These rocks are found on the northwestern coastline of Tenerife in a town called San Juan de la Rambla. There are also very large boulders under the water and at the beach that can be seen while swimming at Playa Buenavista near the famous Masca Gorge.

Cloud Line

Above the cloud line on El Teide at about 2300 m.

In ascending the mountain on a clear day, the cloud line becomes very evident. Behind the small *peak*, the layers of clouds which form the cloud line can be seen. This is the area in the sky at which most of the cloud formation can happen due to *inversion layers*. Inversion layers occur when a mass of warm air floats on top of a mass of cooler air. The cooler layer is then trapped closer to the Earth surface. Normally, the warmer air is closer to the Earth. Hence the name inversion layer. Airplanes usually fly above the cloud line. The volcanic island of Las Palmas can be seen in the distance.

Volcanic Rock and Lava

Lichen growth on mafic rock at 3500 meters

Volcanic rock formations near El Teide peak

Obsidian-glassy extrusive lava type at 3500 m.

Right side: close up of lava

Lava Channels

At Masca Gorge, massive viscous lava channel formations rise 1000 meters high.

In the Masca Gorge region north of Los Gigantes near the northwestern corner of Tenerife, acantilados or cliffs, are the steep vertical rock formations that rise about 800 –1000 meters in elevation. These lava channels formed only 160,000 years ago, according to the latest scientific studies, after the plate shift and when the north side of Tenerife slid into the ocean, forming the Icod Valley. It was near that time the volcano formed, around 120,000 years ago. The Gorge affords an awesomely fun, three-hour, 6 mile/10k vertical hike from the town of Masca to the Playa Buenavista, for the strong of heart, body and mind!

Sun Glint

Sun Glint over the Atlantic Ocean

Sun glint is a strong reflection of the sun on the surface of the ocean. It can be seen from the volcano above cloud line. This visual phenomenon occurs when the angle at which the sun is reflected on the ocean is the same as the angle at which it is being viewed. Satellites often capture sun glint. It can cause difficulties in studying the ocean water column. Could scientists measure the temperature of the water column under the sun glint to determine if there is any increase in temperature due to the constant impact of the sunlight? Would this reflected sunlight, towards the area of hotspot increase the chances of volcanic activity? Viewed from on high, sun glint adds a golden touch to a serene scene. The circular volcanic island in the background is the designated excellent World Biosphere Reserve of La Gomera.

Cable Car

Cable Car Ride to the Upper Station

This cable car on El Teide takes thousands of passengers up onto the volcano in a year. A ticket costs about 27 Euro. El Teide is the most visited tourist destination in Spain. From the lower station base at 2,356 meters it is an eight-minute ride to the upper station. The cable car is a very durable and modern convention for tourism purposes. Edmund Scory was the first to ascend the mountain. When Edmond Scory, in 1582, or Alexander von Humbolt, in June of 1799, climbed El Teide, they had no such advanced technologies or conveniences. Hiking the entire mountain slope on foot or on horseback can take many hours, if not days. Journeying on foot with backpacks and camping gear requires a clear understanding of the high altitude environment.

Upper Station Cable Car House on El Teide

Nowadays, the Upper Station is where you arrive to explore, on foot, the rest of the volcano. There are two hiking routes to take from this area. The route to the summit requires a free permit. The Upper Station house is a very solid, modern construction at 3,555 meters in elevation. Temperature changes are quite evident as it is much colder and windier at the Upper Station than at the Lower Station. Make good preparations for hiking El Teide, as for any other altitude climb, so that you have the strength and tenacity for a short hike from the Upper Station.

Volcano Peak

El Teide Peak

This is the view of El Teide in looking up to the peak from the level of the upper station cable car house. When the clouds clear it is easy to see the outline of the volcano, and view hikers ascending on the left side crest of the mountain. When you are at the top, take at least 40 minutes to look around and walk on the non-permit required foot path. Be very self-aware and monitor your own heart rate and breathing. If you feel lightheaded or dizzy while walking, stop to pause for a while till you acclimate. Know that there can be a 30-minute wait to catch the cable car back down to the lower station house. Enjoy the tremendous view of La Caldera de Las Cañadas, El Chinyero, the peak of El Teide, the different types of lava, and lava rocks. When you arrive at the lower station again, take note of the Balancing Rock.

Activities and Research Projects

A. Fun Geology Activities to Think About and Do

1. Go outside and find interesting rocks. Bring them in and start a collection. Get a notebook, and write down each rock name, take a picture, add characteristics. Over the course of several months or years, develop an interesting collection of rocks that you find in your travels.

2. With the permission of adults, parents or teachers, take a field trip to a geologically interesting location. This can be a volcano, a crater, a caldera, a mesa, Arches National Park, mountains nearby where you live, the water's edge, coastal areas etc. Look for rocks and evidence of geological layers in the earth. Write up an interesting paper with illustrations describing the location and what sort of geological layers you found.

3. Make a shoebox diorama showing an area of the earth that you found to be geologically interesting. Use colors and labels to add detail for understanding of the structure.

B. Different Types of Rocks Found on Volcanoes

Look up famous volcanos and study the rocks that are on them. List the volcanoes along with the rock types. Make lists of these types of rocks. Try to group or classify them.

_____

_____

_____

_____

_____

_____

C. The Major Volcanoes on Earth

Do more research on the internet and in books. Find the 5 largest volcanoes on Earth. Determine the height from sea level to the peak elevation. List the information below.

1._____

2._____

3._____

4._____

5._____

Look up more information on volcanoes. List ten more volcanoes and the regions of the Earth in which they are found.

6._____

7._____

8._____

9._____

10._____

11._____

12._____

13._____

14._____

15._____

Map out the locations of all these volcanoes. Go to the Smithsonian website listed in the Reference section. For each volcano, chart information which includes the name, the height, the continent, the country, the tectonic plate and the last eruption. Determine patterns among the locations to understand how the Earth is acting and reacting when it goes through eruptions. Use GIS software or Google maps, if possible, when locating and mapping the volcanoes. Use Excel to create your chart. Create legends for your maps. Discuss your thoughts and findings in small groups or teams, displaying your detailed maps to communicate your information.

1. Chart of Information on Volcanoes
2. Map of Volcanoes
3. Write Up of Analysis of Volcanic Eruptions using Pattern Identification or Recognition
4. Discussion Groups

Notes

Glossary

Acantilados- Spanish word for cliffs

Boulders- very large fragments of rock usually in the shape of imperfect weathered rectangles or squares. A boulder defined by a geologist is a size fraction larger than 10.1 inches (26 cm)

Caldera- the enormous bowl shape created when an earthquake shifts or volcanic explosion takes place moving the earth. Caldera look like gigantic craters.

Elevation- the height above or below sea level, measured in meters/feet or kilometers/miles.

Fumarolic- An adjective describing a characteristic of a volcano. Fumaroles are vents or openings in the planet's surface from which volcanic gas escapes into the atmosphere. They may occur along tiny cracks or long fissures, in clusters or fields. Fumarolic is a formational feature from degassing of the magma on its way up. Imagine opening up a soda bottle and the gas goes from solution to gas phase.

Decade volcano- is an appreciation of the time between eruptions, approximately on the decade (10 year scale) scale. It there were several hundred years in between, it could be called a century volcano.

Inversion Layers- Inversion layers occur when a mass of warm air floats on top of a mass of cooler air.

Lithosphere- The uppermost layers of the Earth.

Los Gigantes- a town in Tenerife on the northwestern corner of the island. The town faces the volcanic island La Gomera.

Mantel- The layer of the earth below the crust.

Magma- the molten lava which rises out of the earth in volcanic eruptions. Magma/lava. Magma is always inside the crust. As it comes out of the crust, it's called lava. Only a matter of where it is in relation to the surface. The crust is part of the lithosphere.

Peak- the highest point, or high points in the mountainous landscape of the volcano.

Rift- an area where the Earth's crust and lithosphere are being pulled apart forming long deep cavities.

Slope- the sides of the volcano. This area can be alpine with trees, barren, desert like, soil with plants and cacti. At different elevations, the slope will have different vegetation.

Strombolian- Strombolian volcanic eruptions are mildly explosive. They have a VEI -volcanic explosive index of about 2 to 3 volcanic eruptions. They are named after the Italian volcano in Stromboli, Italy.

Vegetation- the plants which can grow in a region of the earth

Volcanic- an adjective which is used in referring to volcanoes

References

1. http://www.volcanodiscovery.com/what-is-a-fumarole.html

2. http://www.rightpronunciation.com/languages/spanish/el-teide-17650.asp?id2=36&page=141#

3. Volcanoes: A Firefly Guide by Mauro Rosi et al, 1st ed, 336 p.

4. https://www.telefericoteide.com/en/national_park/teide_and_science

5. https://en.wikipedia.org/wiki/Alexander_von_Humboldt

6. http://www.sciencedaily.com/releases/2012/04/120413101117.htm

7. http://www.amazon.com/Volcanoes-World-Lee-Siebert/dp/0520268776

8. http://volcano.si.edu/learn_galleries.cfm?p=12

9. http://volcano.si.edu/volcano.cfm?vn=383030

10. http://volcano.si.edu/volcano.cfm?vn=383030

11. http://nmnh-arcgis01.si.edu/thisdynamicplanet/

Photo Credits

All photos are taken by Pia Lord, on her IPHONE 6Plus camera with the exception of the photo of Pia Lord. This is taken by Fredrik Von Weisz.

About the Author

Pia Lord holds an M.S. in Space Science. She attended undergraduate college at Columbia University, Barnard College in NYC. Her studies include lunar geology which is very much the study of craters, varying surface composition and bombardment of a surface. She studied not just Earth's moon, but also the satellite moon of Jupiter, Io, which is highly influenced by volcanic activity and gravitational interactions of the host planet. She has written 13 books in various genres on topics including space science, music and poetry, planetary physics, orbital debris, biological life cycles and sports skiing stories. In writing her books, she likes to collaborate with artists and scientists, as well as her son. Her interests include reading, open water swimming, skiing, playing the piano and singing first soprano in church choir. She lives with her husband and son in New Jersey, United States of America.

Author and scientist, Pia Lord, exploring on the south facing slope of El Teide Volcano in Tenerife, Canary Islands, Spain.

Frederick Von Weisz, geology consultant, hiking on El Teide.

www.ingramcontent.com/pod-product-compliance
Lightning Source LLC
Chambersburg PA
CBHW041304180526
45172CB00003B/967